小凯特的大收藏

【美】芭芭拉·德鲁波提斯◎著
【美】吉奥雅·法蒙吉◎绘
范晓星◎译

天津出版传媒集团
新蕾出版社

图书在版编目（CIP）数据

小凯特的大收藏/（美）德鲁波提斯(deRubertis,B.)著；（美）法蒙吉(Fiammenghi,G.)绘；范晓星译.
—天津：新蕾出版社，2014.1(2024.12重印)
（数学帮帮忙·互动版）
书名原文：A Collection for Kate
ISBN 978-7-5307-5896-0

Ⅰ.①小…
Ⅱ.①德…②法…③范…
Ⅲ.①数学–儿童读物
Ⅳ.①O1-49

中国版本图书馆CIP数据核字(2013)第270450号

A Collection for Kate by Barbara deRubertis;
Illustrated by Gioia Fiammenghi.
Copyright © 1999 by Kane Press, Inc.
All rights reserved, including the right of reproduction in whole or in part in any form. This edition published by arrangement with Kane Press, Inc. New York, NY, represented by Lerner Publishing Group through The ChoiceMaker Korea Co. Agency.
Simplified Chinese translation copyright © 2014 by New Buds Publishing House (Tianjin) Limited Company
ALL RIGHTS RESERVED
本书中文简体版专有出版权经由中华版权代理中心授予新蕾出版社（天津）有限公司。未经许可，不得以任何方式复制或抄袭本书的任何部分。
津图登字：02-2012-215

出版发行	天津出版传媒集团 新蕾出版社
	http://www.newbuds.com.cn
地　　址	天津市和平区西康路35号(300051)
出 版 人	马玉秀
电　　话	总编办(022)23332422 发行部(022)23332679　23332351
传　　真	(022)23332422
经　　销	全国新华书店
印　　刷	天津新华印务有限公司
开　　本	787mm×1092mm　1/16
印　　张	3
版　　次	2014年1月第1版　2024年12月第24次印刷
定　　价	12.00元

著作权所有，请勿擅用本书制作各类出版物，违者必究。
如发现印、装质量问题，影响阅读，请与本社发行部联系调换。
地址：天津市和平区西康路35号
电话：(022)23332351　邮编：300051

无处不在的数学

资深编辑　卢　江

人们常说"兴趣是最好的老师",有了兴趣,学习就会变得轻松愉快。数学对于孩子来说或许有些难,因为比起语文,数学显得枯燥、抽象,不容易理解,孩子往往不那么喜欢。可许多家长都知道,学数学对于孩子的成长和今后的生活有多么重要。不仅数学知识很有用,学习数学过程中获得的数学思想和方法更会影响孩子的一生,因为数学素养是构成人基本素质的一个重要因素。但是,怎样才能让孩子对数学产生兴趣呢?怎样才能激发他们兴致勃勃地去探索数学问题呢?我认为,让孩子读些有趣的书或许是不错的选择。读了这套"数学帮帮忙",我立刻产生了想把它们推荐给教师和家长朋友们的愿望,因为这真是一套会让孩子爱上数学的好书!

这套有趣的图书从美国引进,原出版者是美国资深教育专家。每本书讲述一个孩子们生活中的故事,由故事中出现的问题自然地引入一个数学知识,然后通过运用数学知识解决问题。比如,从帮助外婆整理散落的纽扣引出分类,从为小狗记录藏骨头的地点引出空间方位等等。故事素材全

部来源于孩子们的真实生活,不是童话,不是幻想,而是鲜活的生活实例。正是这些发生在孩子身边的故事,让孩子们懂得,数学无处不在并且非常有用;这些鲜活的实例也使得抽象的概念更易于理解,更容易激发孩子学习数学的兴趣,让他们逐渐爱上数学。这样的教育思想和方法与我国近年来提倡的数学教育理念是十分吻合的!

这是一套适合5~8岁孩子阅读的书,书中的有趣情节和生动的插画可以将抽象的数学问题直观化、形象化,为孩子的思维活动提供具体形象的支持。如果亲子共读的话,家长可以带领孩子推测情节的发展,探讨解决难题的办法,让孩子在愉悦的氛围中学到知识和方法。

值得教师和家长朋友们注意的是,在每本书的后面,出版者还加入了"互动课堂"及"互动练习",一方面通过一些精心设计的活动让孩子巩固新学到的数学知识,进一步体会知识的含义和实际应用;另一方面帮助家长指导孩子阅读,体会故事中数学之外的道理,逐步提升孩子的阅读理解能力。

我相信孩子读过这套书后一定会明白,原来,数学不是烦恼,不是包袱,数学真能帮大忙!

凯特低头坐在座位上，一副愁眉不展的模样。

老师正在最后一次提醒大家："下周是收藏品展示周。如果有谁报名参加了，一定要记得按时带来你们的宝贝！"

凯特偷偷看了一眼日历。她是报名星期四参展,可问题是,她还什么收藏品也没有呢。她只剩下不到一周的时间来准备了。"啊,天哪。"她低声咕哝了一句。

报名是好几个星期前的事了。那天,好多同学争先恐后地在日历上写下自己的名字,所以凯特也报了名。当时看来还有很长的准备时间呢。可现在眼看就到了!

周末的大部分时间里凯特都忙着找收藏品。她把衣橱、抽屉和玩具箱翻了个底朝天,书架最上面、床底下,她都没放过。

　　她这种东西有一点儿,那种东西有一点儿,但哪种东西的数量也没多少。一种东西要有多少才算得上是收藏呢?

　　凯特决定先等一等,看看其他同学都带什么到学校再说。

星期一上午,约瑟夫第一个展示他的收藏。他提来两个大包,里面全是书!他都快提不动了!

"我收藏了很多关于爬行动物的书。"他说。他先给大家展示了9本关于蛇的书,又展示了5本关于蜥蜴的书。凯特马上心算出了书的总数。

唉,怎么办呢?我可没有哪种东西能凑够14件!凯特想。

下面轮到艾玛了,她迫不及待地向同学们展示她的藏品。

"我收藏了两种冰箱贴。"她说。她有 13 个动物冰箱贴,包括长颈鹿和斑马图案,还有 11 个食物冰箱贴,包括比萨饼和果酱夹心曲奇图案。那个曲奇看上去棒极了,叫人忍不住想咬一口。

凯特在纸上写下了 13 和 11。现在她不得不做道两位数的加法！

她已经想到这肯定是大数。

凯特咕哝道:"无论哪种东西我也绝对不可能有 24 件。"

那天晚上，凯特在家里翻箱倒柜。

她找到点儿这个——4本关于马的旧书，又找到点儿那个——5个冰箱贴。可她哪种东西也没找到很多。

星期二,本展示了他的收藏品——贝壳。他有整整三盒贝壳呢。第一个盒里有 15 枚,是他从佛罗里达州带回来的。第二个盒里有 10 枚,是他从加利福尼亚州带回来的。第三个盒里有 5 枚,是他在夏威夷州的奶奶送给他的。

本把各种各样的贝壳全给大家展示了一遍。凯特则飞快地在心里把贝壳的总数算了出来。这些数加起来很简单:15+10=25,25+5=30。

30！大家的收藏品数量变得越来越多了！凯特面临的问题也越来越大了。没错，她也有一些贝壳，但6枚贝壳算不上是收藏啊。

下一个是琼。凯特心想：她总爱显摆，可她今天只带来了一个盒子，也许……

　　"去年夏天，我们全家来了一次很长很长很长的旅行。"琼开始讲起来，"每到一个地方，我都会买几张明信片。我们去的地方很多很多，所以我收藏了很多很多明信片！"

琼数了数明信片。凯特想：她的明信片永远也数不完。

琼总共有 11 张科罗拉多州的明信片、13 张新墨西哥州的、15 张亚利桑那州的。

凯特把这些数都记了下来。嗯……这几个数每个数都大于10，那总数一定大于30了！她琢磨着。

"39！"凯特惊讶极了，只说了句，"啊，天哪！"

那天晚上，凯特从抽屉里翻出 2 张旧明信片，从沙发缝里找到 1 张。爸爸从邮箱里找到 1 张明信片也给了凯特。一共 4 张明信片。

然后，凯特把那6个贝壳放进一个盒子里。只是一小盒。

"啊，天哪！"凯特叹息道。

到了星期三，雷切尔跟大家分享了自己的"猪猪"收藏品。好多小猪啊！连她的妈妈都得来帮忙了，她们搬来了6个鞋盒，里面装的全都是小猪！

盒子里面有 10 只木头小猪、12 只琉璃小猪、16 只金属小猪、4 只塑料小猪和 7 只长毛绒小猪。大家都兴奋地喊:"哇!"雷切尔更是开心得像只快乐的小猪,只有凯特怎么也高兴不起来。

　　凯特又气又烦。雷切尔有那么多小猪,她算总数的时候还要进位呢!

　　她盯着纸上的得数,惊讶地说:"算得对吗?"这是个好大的数呀!也许她算错了。

我还是用计算器再验算一次吧,凯特心想。她用计算器把数全都按了一遍。

49,还是这个数。

"我觉得不舒服。"她叹息道,"明天我可能没法来上学啦……"

那天晚上，凯特看着自己找到的所有东西，这种东西有点儿，那种东西有点儿，哪种也没多少。

她把这些东西一小堆一小堆地摆在床上：

她有4本关于马的书、5个冰箱贴、6个贝壳、4张明信片。

她没有小猪，可是她有3只青蛙和5只泰迪熊。

渐渐地，凯特的脸上露出了笑容。哈哈！我有办法了！她想。

然后，她把床上的东西全都装了起来。

第二天到了学校，凯特把她所有的东西全摆了出来。大家都围上来看，没有一个人说话。

最后，老师开口了："凯特，你收藏的是什么？"

凯特自豪地笑了，她说："我的收藏就是各种各样的大收藏！"

"太棒啦！""真是个好主意！"同学们都赞不绝口。

加法

这里是几种加法运算：

1. 增数法　5+2=？

比5多1是6,比6多1是7。

2. 双倍法　5+6=？

5+5是10,再多1是11。

3. 交换法　2+8=？

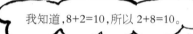

我知道,8+2=10,所以 2+8=10。

4. 间隔计数法

5,10,15,20　　10,20

5. 找规律

　　20+10=30　　　　70+10=80　　　　40+10=50

　　30+20=50　　　　40+50=90　　　　70+30=100

6. 用加法法则做两位数加法竖式

先加个位。再加十位。　　　　　　　逢十进位。

```
    14          1                        1
    23         36          52           85
 +  41       + 48        + 96         + 39
   ────       ────        ────         ────
    78         84         148          124
```

不需要进位　　个位相加后进位　　十位相加后进位　　个位、十位相加后都需要进位

亲爱的家长朋友,请您和孩子一起完成下面这些内容,会有更大的收获哟!

提高阅读能力

- 阅读封面,包括书名、作者等内容。然后和孩子聊聊,我们有可能收藏什么东西?书中的小主人公凯特收藏的又是什么呢?
- 在阅读过程中,让孩子模仿凯特,做做这些动作:愁眉不展、自言自语、翻箱倒柜等。
- 故事中的约瑟夫收藏了 14 本关于爬行动物的书,艾玛收藏了 24 个冰箱贴,请孩子谈一谈自己有什么收藏品。如果没有,问问他打算收藏什么。
- 读完故事,凯特的收藏和你们原先猜想的是否一样?

巩固数学概念

- 在阅读的过程中,和孩子一起检查凯特的加法结果,可以先让孩子心算,再用计算器验算。
- 请利用第 32 页上的内容,分析凯特用了哪些加法运算的方法:增数法、双倍法,还是交换法?
- 仔细欣赏插图,帮助孩子提高读图观察能力。例如,在第 30~31 页,凯特的收藏里每类东西各有几个?一共有几类收藏?凯特的收藏数量分别比雷切尔和约瑟夫的多还是少?
- 请把下面这些与数学知识相关的词语抄在卡片上:两位数、总数、进位、计算器等。让孩子在书中找到这些词语,讨论一下这些词语怎样使用。也可以让孩子通过画图来说明。

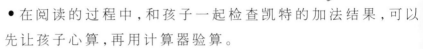

生活中的数学

- 请您和孩子也整理一些"包罗万象"的收藏,让孩子把收藏中每类物品的数量都记录下来。
- 利用第 32 页中加法运算的方法,计算你们收藏物品的总数,鼓励孩子用另一种方法或者计算器来验算结果。

你能把每束花按数量插进相应的花瓶里吗？相信你一定行！

小朋友们，除了第 32 页介绍的几种加法运算方法，我们还可以利用数的组合来计算加法哟！比如：

$$\begin{matrix}5\\ \diagup\diagdown\\ 1\quad 4\end{matrix} \rightarrow \begin{matrix}1+4=5\\ 4+1=5\end{matrix} \qquad \begin{matrix}8\\ \diagup\diagdown\\ 5\quad 3\end{matrix} \rightarrow \begin{matrix}5+3=8\\ 3+5=8\end{matrix}$$

收藏品大拍卖

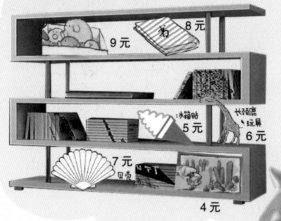

我要一本书和一个长颈鹿玩具,需要投多少硬币呢?

我想要一个冰箱贴和一个贝壳,需要投多少硬币呢?

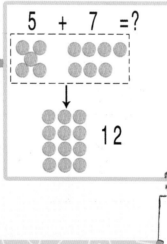

5 + 7 = ?

12

小朋友,你想买什么呢?算算要投多少硬币?

$$\begin{array}{r} 5 \\ +7 \\ \hline 12 \end{array}$$

互动练习

哪只小狗拾飞碟的总成绩最好呢？

参赛者	第一轮	第二轮
	10 个	5 个
	12 个	3 个
	11 个	6 个

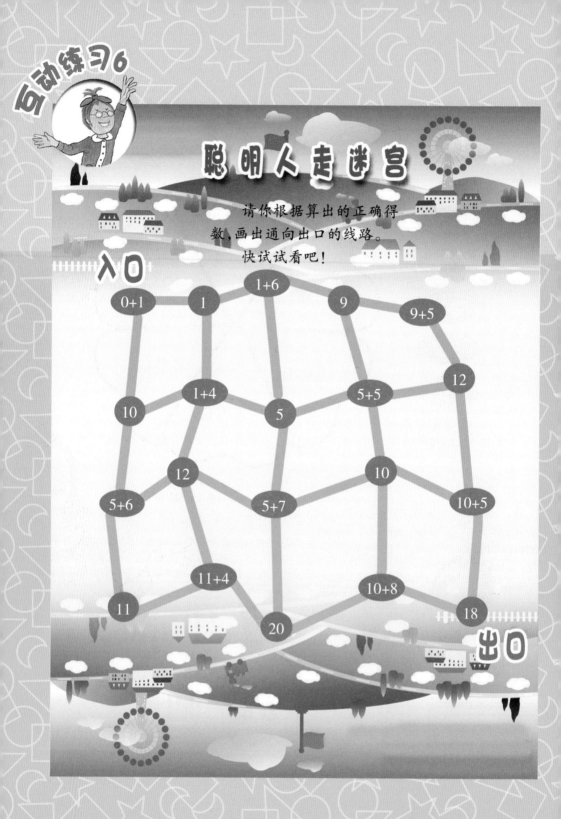

谁转到的数字之和最大,就可以得到大奖!

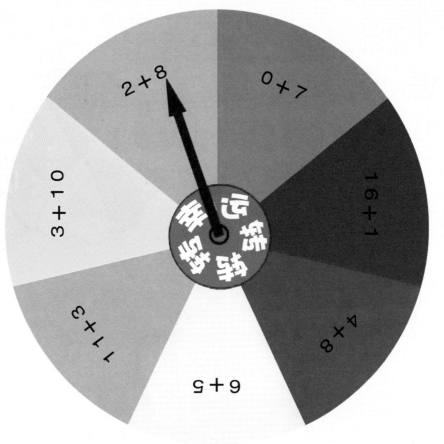

我要转到哪个颜色才能得到大奖呢?

41

互动练习1：

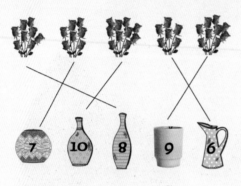

互动练习2：

互动练习3：8+6=14

互动练习4：10+5=15
12+3=15 11+6=17

互动练习5：

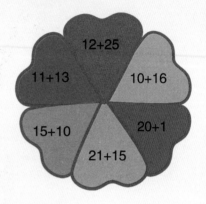

互动练习6：

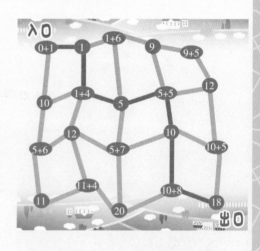

互动练习7：紫色

(习题设计：骆　双)

A Collection for Kate

Kate slumped in her seat. She frowned.

Her teacher was giving the class one last reminder. "Next week is collection week. If you signed up to share your collection, be sure to bring it on the right day!"

Kate peeked at the calendar. She had signed up for Thursday. The problem was that she didn't have a collection. And she had less than a week to get one. "Oh, brother," she muttered under her breath.

Sign-up day had been weeks ago. Lots of kids had rushed to put their names on the calendar, so Kate did too. It had seemed like she would have PLENTY of time to get a collection. Not anymore.

Kate spent most of the weekend looking for a collection. She looked through her closet, her drawers, and her toy box. She looked on top of her bookshelves. She looked under her bed.

Kate had a little of this. She had a little of that. But she didn't have a lot of anything. How many of something did she need to make a collection?

She decided to wait and see what the other kids would bring.

On Monday morning, Joseph showed his collection first. He had two tote bags FULL of books. He could hardly carry them all!

"I collect books about reptiles," he said. First he showed 9 books about snakes. Then he showed 5 books about lizards. Kate added them up in her head.

"Oh, brother!" she thought. "I don't have 14 of anything."

Next came Emma. She could hardly wait to show her collection.

"I collect two kinds of magnets," she said. She had 13 animal magnets, including a giraffe and a zebra. And she had 11 food magnets, including a pizza slice and a cookie with jelly in the middle. The cookie looked good enough to eat.

Kate wrote down the 13 and the 11. Now she had to add 2-digit numbers!

She already knew this was going to be a big sum.

Kate moaned, "I definitely don't have 24 of anything."

That night Kate searched the house.

She found a little of this—four old books about horses. She found a little of that—five magnets on the refrigerator. But she didn't find a lot of anything.

On Tuesday, Ben shared his shell collection. He had three boxes FULL of seashells. The first box held 15 shells from a trip to Florida. The second box held 10 shells from a trip to California. And the third box held 5 shells from his grandmother in Hawaii.

Ben showed all the different kinds of shells. Meanwhile, Kate quickly added up the numbers in her head. They were easy to add: 15 plus 10 equals 25, plus 5 more makes 30.

Thirty! The collections kept getting bigger! And so did Kate's problem.

Sure, she had a few seashells. But six shells were NOT enough for a collection.

Joan was next. "She's always such a show-off," thought Kate. "But she only has one box. Maybe..."

"My family took a loooong trip last summer," Joan began. "I bought postcards at all the places we visited. We went to LOTS of places. So I have LOTS of postcards in my collection."

Joan counted the postcards. Kate thought she would never stop counting.

Joan had 11 postcards from Colorado, 13 from New Mexico, and 15 from Arizona.

Kate wrote down the numbers. "Hmmm...Each of these numbers is bigger than 10," she thought. "So the sum must be more than 30!"

"Thirty-nine!" Kate gulped. "Oh, brother," was all she could say.

That night, Kate found two old postcards in a drawer. She found one in between the sofa cushions. And Dad gave her one that had come in the mail. Four postcards. Big deal.

Then Kate put her 6 seashells in a box. A small box.

"Oh, brother," Kate muttered.

On Wednesday, Rachel shared her pig collection. It was huge! Her mom had to help her carry the six shoe boxes filled with pigs.

The boxes held 10 wood pigs, 12 glass pigs, 16 metal pigs, 4 plastic pigs, and 7 plush pigs. Everyone said, "Wow!" Rachel beamed. But Kate did not.

Kate fussed and fumed. Rachel had so many pigs, Kate had to regroup to find the sum!

She stared at the number. "Can that be right?" she wondered. It was such a BIG number! Maybe she had made a mistake.

"I'd better check this sum on my calculator," Kate thought. She punched in all the numbers.

There it was again—49.

"I feel sick," she groaned. "I might not be able to come to school tomorrow...."

That night Kate looked at all the stuff she'd found. She had a little of this. She had a little of that. But she didn't have a lot of anything.

Kate arranged all the little groups on her bed.

She had 4 books about horses.

She had 5 magnets, 6 shells, and 4 postcards.

She had no pigs.

But she did have 3 frogs and 5 teddy bears.

Slowly, Kate began to smile. "Ah-ha! That's it!" she thought.

Then she quickly packed up everything on the bed.

At school the next day, Kate set out all her things. Everyone watched. No one said a word.

Finally her teacher spoke. "What is it that you collect, Kate?" she asked.

Kate smiled proudly. "I collect COLLECTIONS!" she said.

"Cool!" "What a great idea!" said the kids.

"Very nice, Kate," said her teacher. "I suppose you didn't want to collect just one kind of thing."

"Right," said Kate. "I love the collection of everything!"